C000319107

THE LIFE & TIMES OF

Albert Einstein

BY
James Brown

SHOOTING STAR PRESS

This edition printed for, Shooting Star Press Inc, 230
Fifth Avenue, Suite 1212, New York, NY 10001

Shooting Star Press books are available at special discount
for bulk purchases for sales promotions, premiums, fund-
raising or educational use. Special editions or book
excerpts can also be created to specification. For details
contact – Special Sales Director, Shooting Star Press Inc.,
230 Fifth Avenue, Suite 1212, New York, NY 10001

This edition first published by Parragon Books
Produced by Magpie Books Ltd, 7 Kensington Church
Court, London W8 4SP
Copyright © Parragon Book Service Ltd 1994
Cover picture & illustrations courtesy of: Associated Press;
Rex Features; Mary Evans Picture Library.

This book is sold subject to the condition that it shall not,
by way of trade or otherwise, be lent, resold, hired out or
otherwise circulated without the publisher's prior consent
in any form of binding or cover than that in which it is
published and without similar condition being imposed on
the subsequent purchaser.

ISBN 1 57335 049 4
A copy of the British Library Cataloguing in Publication
Data is available from the British Library.

Typeset by Hewer Text Composition Services, Edinburgh
Printed in Singapore by Printlink International Co.

CHILDHOOD

Nothing in Einstein's family background affords one much clue as to his extraordinary talent. He was born on 14 March 1879 in the city of Ulm in Southern Germany, thus conferring on him a nationality which would prove a persistent problem in subsequent years. His father, Hermann Einstein, was an amiable, gentle business man, with sadly little talent for business. His mother, Pauline Einstein (née Koch), had somewhat more artistic tastes than her husband, being particularly given to reading and music – tastes her son would inherit. She was also the daughter of a reasonably successful grain merchant, which was just as well for Hermann, because the Kochs repeatedly had to help him out.

Albert Einstein was one of two children. His sister, Maja, recalls that Albert was quieter than other children, and would sometimes be called upon to settle their disputes. Others, however, were less impressed by the young Einstein. The rest of his family were somewhat concerned at the time it took him to learn to talk. Having learned, he would practise entire sentences before starting to speak – perhaps an early indication of a preference for living within the world of his own mind.

Others were more openly disparaging about the young Einstein's talents. The Einstein family was Jewish by descent, but they were not practising Jews. When he was five, Einstein was sent to a Catholic elementary school. The school and Einstein were not much impressed by each other. Consulted by Hermann Einstein about young Albert's prospects, the headmaster reportedly prophesied that Einstein junior would never amount to anything.

Einstein's problems had little to do with his being the only Jewish boy in a Catholic school, but more to do with his temperament. In later life the causes he supported indicated a dislike of authoritarianism and convention. This dislike took root early. Much of the school's teaching was by rote. Pupils were expected to accept the authority of the teachers on almost everything. Albert Einstein had scant aptitude for being taught, and the school offered scant opportunity for pupils like himself who preferred to discover things for themselves.

Even at the age of five, his curiosity about the hidden forces of the world had been aroused. He was ill and his father tried to amuse him by showing him a compass. Einstein was fascinated:

'That this needle behaved in such a determined way did not at all fit into the kind of occurrences that could find a place in the unconscious world of concepts. I can still remember – or at least believe I can remember – that this experience made a deep and lasting impression upon me. Something deeply hidden had to be behind things.'

4

Whether this incident occurred in quite the way he recalled or not, early in life Einstein did indeed start to give some sign that discovering the deeply hidden truths behind the world as it appears would be his vocation.

When he was ten he was sent on to the next level of schooling at the Luitpold Gymnasium (i.e. secondary school), which was even more regimented than his last school, and which he disliked even more. However, his education was proceeding – largely outside the Gymnasium. His uncle Jakob, his father's business partner, was an engineer, and lent him a book on algebra, which the young Einstein delighted in. From 1889 onwards, the Einsteins used to have Max Talmud, a poor medical student, round to dinner. Though twice Albert's age, Talmud and

Albert quickly became friends, and Talmud lent him books on science, mathematics and philosophy.

In the midst of these developing secular interests Einstein suddenly had a burst of religious fervour. For about a year when he was twelve, Einstein read the Bible, observed the dietary laws strictly, and composed songs in praise of Judaism which he would sing to himself on his way to school. But he then equally suddenly turned against formal religion, as the strict demands of faith came into conflict with his otherwise sceptical, enquiring temperament.

The continuing sorry state of his father's business ventures induced Hermann and Jakob Einstein to try their fortune in Italy. In 1894 the family went south. But Albert was left in lodgings to complete his education at the Gymnasium. Appalled at the prospect, he deviously secured a doctor's certificate attributing to him a nervous disorder, which was good enough excuse to leave. On presenting it to the school authorities, he was, however, mortified to be told that he was in any case expelled. It is hard to know precisely what he had done wrong. One teacher peevishly complained to him that he somehow undermined the other students' respect for the teacher's authority.

In any case, Einstein followed his family south, where he arrived somewhat to their surprise and announced his rejection of Judaism and German citizenship.

STUDENT DAYS

Einstein told his parents that he intended to apply for a place at the Federal Institute of Technology in Zurich, Switzerland. Meantime he set off on a cultural jaunt around Italy, relishing Italian art and the more relaxed tenor of Italian life.

Einstein was a year younger than most of the other candidates for the Institute. In the entrance exam he did well in maths and science, but the scope of the exam was broad, and overall he did not do well enough to get in.

Einstein had evidently impressed the Principal, who showed an interest in his progress by suggesting that he attend a Swiss secondary school for a year. A family friend recommended the school at Arrau. It was a happy choice. Einstein lodged with the head, Jost Winteler, who became a lifelong friend – indeed, Maja Einstein would later marry Jost Winteler's son, Paul.

His time at Arrau was happy and fruitful. In the autumn of 1896 he secured the diploma which would enable him to go on to the Institute.

He had decided to study maths and physics. But, oddly enough, the course on which he enrolled was a specialist teacher-training course. Given the single-mindedness with which he later pursued research in his spare time, it is hard to believe that he seriously intended to devote himself to school-teaching. But perhaps his choice was guided by prudent paternal advice.

Einstein was not a model student. He was inclined to be casual about attendance at lectures and about clothes. He did work hard, but only at those things that happened to interest him. The rest was cheerfully jettisoned.

Curiously, among the things he shirked was mathematics. Mathematics is central to science. It is virtually its language, and this is especially true in physics. Einstein's own scientific work leans heavily on mathematics, without which he could never have made precise predictions. Einstein later explained that 'it was not clear to me as a young student that access to a more profound knowledge of the basic principles of physics depends on the most intricate mathematical methods. This dawned upon me only gradually after years of independent scientific work.' As one of his maths teachers put it, Einstein was 'a lazy dog who never bothered about mathematics at all'.

Einstein's fecklessness almost endangered his chances of getting a diploma. He was due to sit his final exams in 1900, and was only saved from failure by the kindness of Marcel Grossman, a fellow-student. Grossman was to become a successful mathematician and was laying the ground for his success by taking meticulous lecture notes. All Einstein had to do was mug up Grossman's notes.

Albert Einstein

The heart of the generalized theory of gravitation is expressed in four equations, shown in the accompanying illustration.

$$g_{ik;\ell} = 0; \quad \Gamma_{\underset{-}{+}i}^{\ell} = 0; \quad R_{ik} = 0; \quad \underset{+}{\overset{i s}{\vee}} g_{,s} = 0$$

German lower case G

The equations have the mathematical properties which seem to be required in order to describe the known effects, but they must be tested against observed physical facts before their validity can be absolutely established.

Einstein's Generalized Theory of Gravitation

In later life, Einstein justified his cavalier attitude to the formalities of academic life. It was, he claimed, the way to retain the freedom to indulge his curiosity, which curricula and exams try to stifle: 'It is, in fact, nothing short of a miracle that modern methods of instruction have not yet entirely strangled the holy curiosity of enquiry.'

Einstein's extracurricular activities were not entirely intellectual. Among the students in his year was Mileva Maric. It was unusual enough at the time to find a female student in higher education, but for a woman to read a science subject was especially rare. Mileva was evidently a person of some resolve, to have made it from a family background of modest means in Hungary.

Towards the end of their course, the relationship of Einstein and Mileva became close. They were an apparently ill-assorted pair: she reserved, but capable of some depth of feeling, while he was ebullient and out-going, but already cultivating the emotional detachment which he felt his scientific work required. Mileva Maric was to become the first Mrs Einstein.

PROFESSIONAL FAILURE; SCIENTIFIC SUCCESS

Einstein's talent for rubbing authorities up the wrong way may have something to do with what happened when he came to apply for jobs. In little more than a decade he was to become one of the most sought-after academics of his generation, but right now he was the only graduate of his year to fail to get a regular academic or teaching post. It's a fair guess that an unfavourable reference may have had something to do with it.

Einstein sent letters to institutions and eminent scientists. In secret, lest he offend his son, Hermann Einstein appealed to Prof. Wilhelm Ostwald at Leipzig – but in vain, although Ostwald and Einstein would later become friends. All Einstein got was a short-term teaching post, filling in for someone called up for military service during the summer term of 1901.

There was a bright spot in this: in February 1901 he finally became a Swiss citizen, which he remained until his death.

But there were also pressing reasons for getting a job – any job. In July, Mileva told him she was pregnant.

Einstein's first resolve was to get a job, marry Mileva, and settle down to family life. But there were problems. His mother was utterly opposed to the match, which may have contributed to the delay of the marriage. At any event, they were still unmarried when a daughter, Lieserl, was born in January 1902. She was presumably adopted. No one has been able to discover what became of her.

Einstein's next temporary job proved to be more temporary than he had intended. He had taken a post for a year as private tutor in a school in Schaffhausen, but for some reason he left after just a term.

Marcel Grossman, whose lecture notes had been so useful, came to Einstein's rescue again. He used his contacts to get Einstein an interview for a post at the Patent Office in Bern. The interview went well. On the strength of it Einstein moved to Bern in

Einstein with other Nobel Prize-winners

experiments. It was this that made it possible for a civil servant without a laboratory or the support of an academic institution to make fundamental advances in physics.

Einstein was not to be without these things for much longer. Early in 1908 he took the first step towards an academic career by becoming *Privatdozent* at the University of Bern. This entailed the delivery of lectures, but did not carry a salary. The idea was that the lecturer be paid directly with fees from his students. Einstein had to fit his lectures around his job, and he gave them between 7 and 8 in the morning. He started with four students, and ended with two: his sister and a close friend, Michelangelo Besso.

While Einstein was expounding 'The Theory of Radiation' to an all but empty room, his name was getting known in scientific circles. In July 1909 he received the first of many honorary degrees – this one from the University of Geneva. Meanwhile efforts were being made to secure for him a newly created post in theoretical physics at Zurich University. On 6 July 1909 he resigned from the Patent Office. For the rest of his life he would be a professional academic.

THE WANDERING PHYSICIST

Einstein's appointment was delayed while the post was offered to Friedrich Adler. The selection board were mostly social democrats, and Adler was the son of the founder of Austria's Social Democratic Party. They reckoned without Adler's punctilious honesty. As soon as he perceived the committee's motives, he withdrew in Einstein's favour.

Einstein became an associate professor, which entailed significant teaching commitments. Being new to this, Einstein found it so time-consuming that he complained to Besso, 'my *real* time is less than at Bern'.

After just a term at Zurich, Einstein was sounded out about the possibility of becoming a full professor at the German University of Prague. He was tempted. However, there were more internal politics to be negotiated before the post fell to him. Prague was then in the Austro-Hungarian Empire. Einstein was German by birth and Swiss by choice. The committee, in defiance of their chairman but in compliance with the Imperial authorities, decided to appoint an Austro-Hungarian, Gustav Jaumann. Fortunately for Einstein, Jaumann was touchy. When he discovered the reason for his appointment,

Jaumann insisted that either the university appoint him on the grounds that he was the finer physicist (which he dogmatically believed) or he would go elsewhere. He went elsewhere.

Einstein moved to Prague in March 1911. As a full professor he was better off than in Zurich, and had greater resources for research. And, in some ways, life in Prague was pleasant for him: he could play the violin in an amateur quartet, and he made various acquaintances, including the writer Franz Kafka.

In other respects, life in Prague was awkward. The German-speaking community made up just 5 per cent of the population and tended to look down its nose at the majority Czech community. Also, though

Einstein had no orthodox faith, Imperial rules required that he declare some religious affiliation. He chose to be classified as a Jew. The Jewish community was caught between the prejudice of both communities.

Einstein's pursuit of professional advancement put a strain on his marriage. By now he had two sons: Hans Albert (born in 1904) and Eduard (born in 1910). The latter was to suffer a lifetime of mental illness. It is clear that Mileva was not happy with the move to Prague. It was some relief to her that Einstein's wanderings soon took him back to Zurich. In 1912 he took up a chair at the Institute where he had been a student.

However, Mileva's relief was destined to be short-lived. Einstein had scarcely got down to work with his old friend Marcel Gross-

At home in Caputh, Berlin

arrangements had deteriorated since his separation from Mileva. Though emotionally somewhat remote, Einstein ideally needed domestic stability – needed, in fact, to be looked after.

Elsa was a cousin of Einstein's. She was a divorcee with two daughters, Ilse and Margot. Perhaps she regarded it as matter of family loyalty that she should nurse her celebrated cousin back to health. But as the relationship developed, they began to think of marriage, and Einstein began to think of divorce. Mileva and Einstein agreed terms a few months after the armistice of 1918.

FAME

Among the casualties of World War One was freedom of communication between intellectuals. This was a grievous loss for the scientific community which was, in many respects, naturally international. But the war obstructed Einstein's work in a more particular way. Einstein was not much given to experiments. However, his theory of relativity did make a number of predictions which could, in principle, be tested. Other scientists were ready to do the hands-on work, and Einstein was naturally eager to see it done.

An opportunity arose when the astronomer Erwin Finlay-Freundlich proposed observing starlight during a solar eclipse. Einstein had predicted that light could be bent by a sufficiently strong gravitational field. In a solar eclipse, stars on the far side of the Sun would become visible from the Earth, and the beams of starlight would pass close to the huge mass of the Sun. If Einstein was right, the apparent position of such stars ought to be slightly different from what was normally observed.

Finlay-Freundlich saw his chance. The next suitable eclipse would occur in the summer of 1914, and southern Russia would be the place from which to view it. With Einstein's help, funds were raised for the trip and the expedition departed.

At this point, World War One broke out. Instead of checking Einstein's theory, Finlay-Freundlich and his team were interned in Odessa. They were later exchanged for a group of captured Russian officers, and trailed home to Berlin.

As it happens, this was a stroke of luck for Einstein, but it must have seemed to be a uniquely well-disguised one at the time. Einstein continued to work on the General Theory during the war, refining and correcting his earlier work. In 1914, though Einstein was correct in predicting a deflection, the numerical value he put on it was just half of what Finlay-Freundlich would have found had he not been stopped by the war. But as the war continued, Einstein had second thoughts and doubled the predicted value of the deflection.

While the world ran mad with blood-lust there was no prospect of testing the revised theory. However, a group of British scientists spotted another opportunity. An eclipse would occur on 29 May 1919. Fortunately the war ended in time for the astronomer Arthur Eddington to assemble an expedition. He was going to go to the island of Principe. Just in case the weather was too bad for clear observation, another expedition was dispatched to Sobral in Brazil. As it happened, both teams were able to make good observations.

There was some delay in announcing the results. Eddington wanted to make his calculations, check them over, and get back to Britain. The formal announcement was finally made in London at a joint session of the Royal Society and the Royal Astronomical Society on Thursday, 6 November 1919: Einstein was right. After standing unchallenged for two centuries, Newton's model of the universe had to be revised.

Headlines the next day flashed the news to the public: the cosmos they inhabited was stranger than they had supposed. It was a cosmos in which space and time were not distinct entities, but were so intricately interrelated that one should speak of the single entity, space-time. It was a cosmos in which matter and energy revealed their unsuspected equivalence, summed up in the most famous scientific formula of the century and possibly of all time: $E=mc^2$, where 'E' is energy, 'm' is mass and 'c' is the speed of light. Since light travels at approximately 186,000 miles a second, and the superscript '2' multiplies that velocity by itself, one does not need much mass to produce staggering quantities of energy. Suddenly the apparently stable, solid world we live in presents itself to the layman's bewildered imagination as huge quanta of energy frozen into the forms of matter. In fact, there is no simple way to convert mass directly into energy – but in nuclear explosions, modest quantities of mass are abruptly transformed into energy.

We revere an Einstein because his work is so far beyond our ordinary concerns. One's reaction to it has an almost religious dimension. In the wake of World War One there was every reason for the public to yearn to be distracted from the ordinary, human world.

35

Einstein's house in Princeton

sense, Einstein himself went along and chortled through the entire proceedings from a box. Anti-Semitism, and its particular manifestation in anti-Einsteinism, could not be treated as a joke for much longer.

Nor was Einstein entirely free of prejudice abroad. There was still some anti-German feeling ready to attach itself, no matter how absurdly, to Einstein. A powerful faction in the Royal Astronomical Society blocked the award of its Gold Medal to him for a while. German scientists were excluded from the first two Solvay Congresses to convene after the war. Admittedly Einstein was exempt from the exclusion, but he felt obliged to stay away in protest.

As if his public problems weren't enough, Einstein had cause for private grief. In

Chaim Weizmann becomes the first President of Israel

Einstein in his garden

January 1920 his mother arrived at his home in Berlin; she was gravely ill, and announced her intention to die. It took her until March to do so, by which time Einstein felt, as he put it, 'completely exhausted'.

It was possibly something of a relief for Einstein that he was much in demand abroad. Indeed, he spent a good portion of his remaining time with Berlin University abroad.

Except for one year in his childhood, Einstein's sense of his own Jewishness had never been pronounced. He was certainly not an orthodox believer. To the extent that he believed in God, his was not the personal God of the Old Testament, but more the God of the seventeenth-century Jewish philosopher, Baruch Spinoza – a God coextensive, in some sense identical, with his own creation. Spinoza was branded a heretic for his views and was banished from the Jewish community in Amsterdam. Einstein distrusted nationalism in any form but the anti-Semitism he encountered fostered a closer identification with his fellow Jews. Chaim Weizmann was an ardent Zionist who would later become the first President of Israel when a Jewish state was finally re-established. When he approached Einstein asking him to join a fund-raising tour of the USA, Einstein agreed.

Weizmann knew Einstein would be a crowd-puller. Even so, when they arrived in New York in the spring of 1921 Einstein was taken aback by the level of press interest. He scored a great success, delivering lectures on relativity (in German with simultaneous

English translation), and making an appeal for funds for a Hebrew University – a cause to which Einstein could unaffectedly lend his whole-hearted support.

Einstein proceeded to Britain, where he had been invited to deliver the Adamson Lecture at Manchester University. Einstein travelled on to London and gave further lectures. Though the press was less strident than its counterpart in the States, coverage of Einstein's visit was intensive.

Back home in Germany the situation was getting uglier. The Foreign Minister of the recently formed Weimar Republic, Walter Rathenau, was assassinated in 1922 simply because he was a Jew. Other Jews in public life were also attacked.

Not surprisingly, given the situation in Germany, Einstein was soon on his travels again – this time to Japan in the winter of 1922. After the furore that greeted him in the USA, he found the calm courtesy of the Japanese more to his taste.

While there, Einstein learned that he was to be the Nobel Laureate in Physics. This was the occasion of further friction with his country of residence. Einstein was unable to collect the award in person, so the ambassador of his State would represent him. But which was that to be: the German or the Swiss envoy? So far as the German authorities were concerned, Einstein had resumed German citizenship on becoming a member of the Prussian Academy of Sciences. This was news to Einstein. Nevertheless, a compromise was thrashed out: the

German ambassador collected the prize in Sweden, but the Swiss ambassador then passed it on to Einstein in Germany.

It would be untrue to suppose that Einstein was universally loathed in Germany. But the people of goodwill who supported him, or at any rate were not prejudiced against him, were apt to remain quiescent. Max Planck, who had been instrumental in securing Einstein for Berlin University, was an honourable exception. Even when others did attempt a supportive gesture, it could misfire. For example, to mark Einstein's half-century in 1929 the Berlin city council was persuaded to give him a house near the river Havel. Einstein loved sailing, so it seemed an appropriate gift. However, the Einsteins discovered that the house was already occupied by people who had no intention of moving out. So the council made another offer: a plot of land, on which Einstein could build his own house. He selected a house near Caputh. But Einstein ended up delving into his savings to pay for the building, which robbed the council's generosity of much of its savour.

As if these problems were not enough to contend with, Einstein's health was not good. On a visit to Switzerland early in 1928, he arrived at a railway station earlier than his host expected, and set off on foot with his bags, only to collapse. It took him months to recover – though the illness did have the pleasant side-effect that it resulted in Einstein engaging a secretary to help him with his correspondence. Helen Dukas would remain in Einstein's service for the rest of his life.

Einstein explains his theories

Princeton University

In any case, Einstein would not have long to enjoy his new home. Regular commitments abroad occupied more of his time. From the end of 1930, Einstein was Visiting Professor at the California Institute of Technology, which would take him to Pasadena for three months each year. In 1931 he was the Rhodes Lecturer in Oxford, and this was followed by a research fellowship at Christ Church. For the next couple of years Einstein spent much time outside Germany, and when in Germany he tended to work in the seclusion of Caputh.

However, this pattern was not to last for long. In December 1932 the Einsteins set off from Caputh for California for their annual visit. As they left their new home, Einstein warned Elsa that they would never see it again. He was right. On 30 January 1933 Adolf Hitler came to power. Einstein never set foot in Germany again.

EXILE

For all his premonitions, Einstein had no way of knowing that he had left Germany for ever. He had laid no long-term plans for living abroad. The next few years would be unsettled.

He had commitments in the USA early in 1933, but then returned to Europe. Prudently, though, he went to Belgium rather than Germany, with which he began to sever his ties. He resigned from the Prussian and

Bavarian Academies, and for the second and last time resigned his German citizenship. The Nazis, not to be outdone, cancelled his now non-existent citizenship.

He passed the summer and autumn of 1933 in Oxford and Belgium. Meanwhile the Nazis continued to vent their spleen by trying to discredit him internationally.

In a sense, the Nazis were right to identify Einstein as a particular enemy. It is hard to imagine a figure more at odds with the ethos of their movement. They were efficiently organized, but deeply and wickedly irrational and destructive. Einstein had such a hazy idea of how to run his life that it is said that he once had to telephone Elsa to ask her, 'Where am I and what am I meant to be doing?' But his mind was deeply rational and creative – possessed by a profound sense of the mystery of things. While the Nazis liked to clothe their inadequacies in handsome uniforms, Einstein's clothes were a shambles.

Einstein's disinclination to dress the part of the great man led to the occasional embarrassment. He had become friends with King Albert and Queen Elisabeth of Belgium while attending the Solvay Congresses. In the midst of this uncertain period, he set off to see them at Laeken. He got lost and slipped into a bar, announced his intention to get to Queen Elisabeth, and went off to wrestle with the mysteries of the telephone. The bartender sized up this disreputably dressed individual and called the police.

By this time, Einstein was widely known for his political views. He supported Zionism and was a pacifist. However, his experience of the Nazis, while it strengthened his Zionism, made him question his pacifism. He arrived at the uncomfortable conclusion that the Nazi threat needed to be countered by force. The pacifist movement was dismayed. But one should note that Einstein's ability to assess scientific problems at speed served him well in this political one. The British government, by contrast, would seek to appease the Nazis until the eve of war itself.

The Nazis indulged themselves by raiding Einstein's house at Caputh, supposedly in search of a communist arms dump. They burned Einstein's book on relativity, and froze his assets.

Though Einstein was out of a job, he was not without prospects. He still held his research

A nuclear explosion, much feared by Einstein

Einstein on his 74th birthday

fellowship in Oxford. He was also being
sought by Abraham Flexner, a persistent
individual, who aimed to establish what
became the Institute of Advanced Study at
Princeton: a place for scientists to work with
each other, free of the burden of regular
teaching commitments. Flexner had had
meetings with Einstein in 1932 in Oxford
and at Caputh, and had struck a deal to
secure his services for part of each year.
He invited Einstein to name his own
terms, only to be taken aback when Einstein
first suggested the absurdly low figure of
$3,000 p.a. and then wondered whether
he could get by on less. Flexner was having
none of this. He went off to negotiate with
Elsa instead and they settled on $16,000.

In October 1933 the Einsteins set sail for the
States. Einstein seems to have intended to

Bust of Einstein by Jacob Epstein

Einstein's principal recourse was to plunge back into his work.

Einstein's secretary, Helen Dukas, now increasingly assumed the duties of housekeeper. His household comprised Helen Dukas, his step-daughter Margot, and his sister, who came to live with him on the death of her husband.

Einstein's growing isolation was professional as well as personal. The most famous scientist of his day was increasingly marginalized within the scientific community because of the nature of his work. Quantum mechanics was being rapidly developed in the early years of this century. It sought to give some account of events on the subatomic scale. At this level, our common-sense understanding of matter breaks down. In the world as we directly perceive it, there is a clear distinction between waves and matter. If one throws a pebble into a pond, for example, one can see waves moving outwards from the spot where the pebble fell. However, that does not mean that the pond itself is moving outwards from the same point. The pond stays put: it is merely the medium in which the waves move. But according to quantum theory, this distinction breaks down when one considers events on a minute scale. Light, for example, has to be considered in some respects as a wave, but in others as a stream of particles.

This in itself was not a problem for Einstein, who had himself been instrumental in developing some aspects of quantum theory. But as the theory developed, it took a form repugnant to Einstein's intuitions about the nature of the universe. According to the uncertainty principle formulated by Heisenberg in 1927, quantum theory puts inherent limits on the extent to which subatomic processes can be measured, and therefore on the extent to which they can definitely be known. When one tries to measure, say, the position and momentum of a subatomic particle, quantum theory insists on the impossibility of knowing both facts simultaneously. In any case, the particle is not only a particle; it is also a wave. And worse still, from Einstein's point of view, was the way in which quantum theory postulates an uncanny interaction between the subatomic objects or events being observed and the observer, so that it is as if the particle/waves know they are being watched and react to it. As a result, our knowledge ceases to be specific and particular, and becomes a question of the probable behaviour of large numbers of such things.

Einstein could not live with this. He wrote in a letter to a colleague:

'Quantum mechanics is certainly imposing. But an inner voice tells me that it is not yet the real thing. The theory says a lot, but does not really bring us any closer to the secret of the "old one" [Einstein's half-joking term for God]. I, at any rate, am convinced that *He* is not playing at dice.'

As he said, on another occasion, 'God is subtle, but he is not malicious'.

The problem was that quantum mechanics was and is a staggeringly successful theory. In blunt terms: it works. As well as trying to pick it apart, Einstein also sought to formulate a unified field theory – a way of conceiving the universe which would subsume the insights of relativity and quantum theory in a single theory. He laboured away at this throughout his time in the USA, but without success – although, more recently some have come round to the idea that such a theory might be possible. In its current form it involves the mathematical postulation of a ten-dimensional universe: one dimension of time and nine of space, six of which are, so to speak, furled round upon themselves, leaving the three of which we have direct experience.

Einstein may have been professionally isolated, but his name still carried enormous weight in the world at large. When fellow physicists realized that the energies locked up in the heart of the atom could in principle be used to create a weapon of terrifying power, they sought out Einstein in 1939, not to help them make the bomb, but to write on their behalf to Roosevelt, warning of the potential of such a weapon and of the possibility that the Nazis might get there first. The latter suspicion proved unfounded, but it impelled Einstein to write to the President. As a result, the Manhattan Project was set up and the USA became the world's first nuclear power. To Einstein's dismay, the bomb was used against Japan. However, Einstein played no active part in the bomb's development. He would probably have found it morally re-pugnant to help make so hideous a weapon, and in any case the FBI had him marked down as a potential subversive. His official work was confined to checking over plans of new systems for the US Navy. His unofficial work consisted in his tireless efforts to help friends and acquaintances escape from the Nazi tyranny.

Einstein retired in 1945, though the pattern of his life did not change much. He still had a room at the Institute and a salary. He still pottered around Princeton, where he was an object of local affection and pride, and the focus of an increasing number of anecdotes. There was the schoolgirl who had problems with her maths, and, remembering that the old man down the road was said to be very good at maths, used to pop round to Einstein's to get his help, which was where her embarrassed mother found her one evening. Or there was the occasion when the secretary at the Institute took a phone call from someone asking for Einstein's address. She explained that she wasn't allowed to give it

1918

Einstein and Mileva agree to a divorce, so he is free to marry Elsa.

1919

It is formally announced that Einstein is right at a joint session of the Royal Astronomical Society.

1920

The Study Group of German Natural Philosophers is established to counter Einstein's theories. Einstein's mother dies in March.

1921

Einstein tours the USA with Weizmann, continuing on to Britain.

1922

Einstein travels to Japan. He learns that he is to be the Nobel Laureate in Physics.

1929

Einstein moves to Caputh.

1930

Einstein becomes Visiting Professor at the California Institute of Technology.

1931

Einstein is the Rhodes Lecturer in Oxford, a research fellowship at Christ Church soon follows.

1932

The Einsteins leave for California never to return to Germany.

1933

The Nazis try to discredit him, and raid his home at Caputh. The Einsteins set sail for the States, Princeton is to become his home.

1934

Ilse, Einsteins step-daughter, dies.

1936

Einsteins wife Elsa dies.

1939

Einstein writes to Roosevelt, warning of the potential danger of the atomic bomb.

1945

Einstein retires, though continues his life much the same.

1952

Einstein is offered the presidency of Israel. He declines.

1955

Albert dies peacefully on 18 April.

LIFE AND TIMES

Julius Caesar
Hitler
Monet
Van Gogh
Beethoven
Mozart
Mother Teresa
Florence Nightingale
Anne Frank
Napoleon

LIFE AND TIMES

JFK
Martin Luther King
Marco Polo
Christopher Columbus
Stalin
William Shakespeare
Oscar Wilde
Castro
Gandhi
Einstein

FURTHER MINI SERIES
INCLUDE

ILLUSTRATED POETS

Robert Burns
Shakespeare
Oscar Wilde
Emily Dickinson
Christina Rossetti
Shakespeare's Love Sonnets

FURTHER MINI SERIES INCLUDE

HEROES OF THE WILD WEST

General Custer
Butch Cassidy and the Sundance Kid
Billy the Kid
Annie Oakley
Buffalo Bill
Geronimo
Wyatt Earp
Doc Holliday
Sitting Bull
Jesse James

FURTHER MINI SERIES INCLUDE

THEY DIED TOO YOUNG

Elvis
James Dean
Buddy Holly
Jimi Hendrix
Sid Vicious
Marc Bolan
Ayrton Senna
Marilyn Monroe
Jim Morrison

THEY DIED TOO YOUNG

Malcolm X
Kurt Cobain
River Phoenix
John Lennon
Glenn Miller
Isadora Duncan
Rudolph Valentino
Freddie Mercury
Bob Marley